Time Motion Studies: A Primer

An AI-Assisted Book by Ted Arehart

Published June 2024

Paperback Version

1st Edition

Table of Contents

Preface 5
Chapter 1: Introduction to Time Motion Studies 9
Chapter 2: Planning Your Study 19
Chapter 3: Data Collection 26
Chapter 4: Data Analysis 34
Chapter 5: Implementing Improvements 41
Chapter 6: Case Studies 49
Glossary 66
Index 70
Appendix 73

"Ted, you really think I can double production next quarter? I can't find welders."

"I think you can septuple production next month. Your welders aren't the problem. It's the fact that they aren't welding."

We look out his window over the manufacturing floor, and his lead welder sets down his welding gun and picks up a broom.

Preface

Who Should Read This Book

This book is for managers, analysts, and anyone involved in optimizing operational efficiency. Whether you are in manufacturing, healthcare, logistics, or any field that benefits from process improvement, this guide is a primer for what is needed to conduct effective time motion studies.

Who Am I?

I have spent the last 7 years in various manufacturing environments conducting time motion studies. My firm, TED Consulting, has assisted companies as large as Ford and Boeing find inefficiencies in their processes, leading to significant productive gains and cost savings. We have also used these techniques for a one-man manufacturing plant to reduce the daily workload from 20-hour days to 8-hour days.

I am an Officer on the Board of Directors of the International MODAPTS® Association as well as an Independent Certified MODAPTS® Instructor. While MODAPTS® is my preferred method for executing a time motion study, other methods can be just as effective depending on the use case.

Why I Wrote This Book

Time motion studies have been a cornerstone of process optimization for decades. With the rise of data analytics and lean methodologies, understanding the intricacies of these studies has never been more critical. This guide aims to demystify the process, providing clear steps and practical advice.

How I Wrote This Book

I wrote this book with significant assistance from a language learning model, specifically Chat GPT-4o. I have reviewed and added to the content provided by the model. While I cannot verify the individual case studies and examples provided, the premise of each is sound, though the purported results are more muted than in my experience.

The quotes at the beginning of each chapter are paraphrasings of my own experience, and not written by AI in any way.

How to Use This Book

I designed each chapter to build on the previous one, taking you from the basics to advanced techniques. Feel free to skip to sections that are most relevant to your needs. The glossary has some of the more technical terms defined. The appendices include templates to help you get started quickly.

*Paperback edition note: To assist with legibility, I have added blank pages or large spaces to improve the reading experience – paraphrasings on the left of their relevant chapters, bullet points wholly on a page, sections appropriately split, etc. Feel free to use these blank areas for notes.

How to Contact Me

If you have questions about this book, want to discuss time motion studies, or need additional resources, please email me: ted@tedconsulting.net.

"So, how do you normally do this?"

"Go out on the line, introduce yourself, and stare at them for 4 hours."

"Seriously?"

"There is no better way to assess a process than through observation. And you don't truly observe unless you can smell the cutting oil."

Chapter 1: Introduction to Time Motion Studies

What is a Time Motion Study?

A time motion study is a business efficiency technique combining time study work (to measure the time taken to complete tasks) and motion study work (to analyze the motions involved in the tasks). It requires breaking down a process into individual steps (obtain bolt, walk to unit, hand start bolt) and allow for categorization and analysis of each piece of a process. A time motion study answers questions like, "How long should it take to do this process?" and "Which parts of this process are a concern?"

History and Evolution

Early Beginnings

- **Frederick Winslow Taylor**: Often referred to as the father of scientific management, Taylor introduced the concept of time studies in the late 19th century. He focused on breaking down tasks into smaller components and measuring each part to improve efficiency.

- **Frank and Lillian Gilbreth:** Pioneers in motion study, the Gilbreths aimed to reduce unnecessary motions in tasks to enhance efficiency and reduce worker fatigue.

Development Over the Decades

- **1920s - 1950s:** The integration of time studies and motion studies into broader industrial engineering practices. The emphasis was on standardizing work processes and improving labor productivity.

- **1960s - 1980s:** The rise of Lean Manufacturing and Total Quality Management (TQM) incorporated time motion studies into larger frameworks aimed at waste reduction and continuous improvement.

- **1990s - Present:** Advancements in technology have transformed time motion studies with tools such as computer simulations, wearable devices, and sophisticated data analytics.

Importance and Benefits

Productivity Improvement

- **Enhanced Efficiency**: By identifying and eliminating unnecessary steps, time motion studies streamline workflows, reducing the time required to complete tasks.
  - **Example**: A manufacturing plant identifies redundant handling steps in the assembly process, leading to a 20% reduction in cycle time.

- **Increased Output**: More efficient processes result in higher production rates without additional resources.
  - **Example**: A retail warehouse reconfigures its layout based on time motion studies, increasing order fulfillment rates by 30%.

Cost Reduction

- **Labor Cost Savings**: By optimizing tasks and reducing wasted effort, companies can achieve significant labor cost savings.

- **Example**: A hospital reduces the time nurses spend on paperwork through default options for the most common issues, allowing them to spend more time on patient care, reducing the need for additional staffing.

- **Resource Optimization**: Efficient use of materials and equipment reduces waste and associated costs.
  - **Example**: A construction company minimizes material handling waste by identifying better routes both on and off site, saving on both indirect material costs and labor hours.

Quality Improvement

- **Consistency and Standardization**: Standardizing processes based on time motion study findings leads to more consistent and higher-quality outputs.
  - **Example**: A food processing plant standardizes its packaging process, resulting in fewer defects and higher product quality.

- **Error Reduction**: Simplifying and optimizing tasks reduces the likelihood of human error.

- **Example**: A customer service center implements standardized scripts and workflows, reducing call handling errors and improving customer satisfaction.

Worker Well-being

- **Ergonomic Improvements**: Time motion studies often highlight ergonomic issues, leading to interventions that reduce physical strain and improve worker comfort.
  - **Example**: An office reconfigures workstations based on ergonomic assessments, reducing instances of repetitive strain injuries.
- **Job Satisfaction**: Efficient processes can lead to more manageable workloads and less frustration for workers.
  - **Example**: A manufacturing team adopts a new workflow that reduces bottlenecks, leading to higher job satisfaction and lower turnover rates.

Key Concepts

Time Study

- **Definition**: A method of measuring the time taken to complete specific tasks using tools such as stopwatches or time-tracking software.
- **Components**:
    - **Task Breakdown**: Dividing work into individual tasks or elements.
    - **Timing**: Measuring the time required to perform each task
    - **Standard Time Calculation**: Determining the average time taken and adjusting for factors like worker fatigue and machine downtime.

Motion Study

- **Definition**: The analysis of the movements required to complete a task, aiming to improve efficiency by eliminating unnecessary motions.
- **Components**:
    - **Task Breakdown:** Dividing work into individual tasks or elements.

- **Motion Analysis**: Studying and optimizing each motion to reduce effort and time.

Time Motion Study

- **Definition**: The method of measuring the time it takes to complete a specific task through the analysis of the movements required to complete that task, aiming to improve efficiency by eliminating unnecessary motions.
- **Components**:
    - **Task Breakdown:** Dividing work into individual tasks or elements.
    - **Motion Analysis**: Studying and categorizing each motion using a Predetermined Time Motion System (PMTS) such as Methods-Time Measurement (MTM), Maynard Operation Sequence Technique (MOST), or Modular Arrangement of Predetermined Time Studies (MODAPTS®).

Integration with Other Methods

- **Lean Manufacturing**: Using time motion studies to support Lean principles such as value stream mapping and waste reduction.

- **Example**: A Lean workshop uses time motion studies to identify and eliminate non-value-added activities in a production line.
- **Six Sigma**: Incorporating time motion studies into Six Sigma projects to identify root causes of inefficiencies and improve process capability.
  - **Example**: A Six Sigma team conducts a time motion study as part of a DMAIC (Define, Measure, Analyze, Improve, Control) project to reduce cycle time in a service process.

"How long do you need?"

"I can do it myself, which will take longer and delay implementation, but spread your cost over time. Or I can put a team together"

"Put a team together"

Chapter 2: Planning Your Study

Defining Objectives

Setting Clear Goals

The success of a time motion study begins with well-defined objectives. Goals should be specific, measurable, achievable, relevant, and time-bound (SMART). Ask yourself:

- What are you trying to achieve?
- Are you looking to reduce cycle time, improve productivity, or eliminate waste?

Identifying Key Performance Indicators (KPIs)

Select KPIs that will help measure the success of your study. Common KPIs include:

- Cycle time: The total time taken to complete a process from start to finish.
- Throughput: The number of units produced, or tasks completed in a given period.
- Efficiency: The ratio of productive time to total time.
- Utilization: The degree to which resources (people, machines) are used.

Selecting the Process

Criteria for Selection

Choose a process that aligns with your objectives and offers significant improvement potential. Consider:

- Frequency: How often is the process performed?
- Impact: What is the effect of the process on overall operations?
- Complexity: Is the process complex enough to benefit from detailed analysis?

Scope and Boundaries

Define the boundaries of your study to keep it manageable. Determine:

- Start and end points of the process.
- Inclusions and exclusions (e.g., specific tasks or sub-processes).
- The level of detail required (e.g., high-level overview vs. detailed task analysis).

Assembling Your Team

Roles and Responsibilities

A successful time motion study requires collaborative effort. Key roles include:

- Project Manager: Oversees the study and ensures objectives are met.
- Observers: Collect data through direct observation or video recordings.
- Analysts: Interpret data and identify improvement opportunities.
- Subject Matter Experts (SMEs): Provide insights into the studied process. These members are critical as they will be able to differentiate waste from value.
- Workers: The individuals performing the studied tasks; their cooperation is crucial. Any implementation plan will fail without their buy-in.

Training Requirements

Ensure all team members understand the objectives, methodology, and tools used in the study. Training may include:

- Time study techniques: Using stopwatches, video recordings, and software.
- Motion study techniques: Identifying and categorizing different motions.
- Time motion study techniques: Certification in chosen PMTS.
- Data analysis methods: Statistical analysis, process mapping, and identifying waste.

Developing a Study Plan

Timeline and Milestones

Create a detailed timeline with key milestones to keep the study on track. Consider:

- Initial planning and preparation.
- Data collection period.
- Data analysis and interpretation.
- Implementation of improvements.
- Post-implementation review.

Resource Allocation

Identify the resources needed for the study, including:

- Personnel: Observers, analysts, and support staff.
- Equipment: Stopwatches, video cameras, computers, and software.
- Budget: Costs for training, equipment, and potential disruptions to operations.

Communication Plan

Effective communication is essential for a smooth study. Develop a plan that includes:

- Regular updates to stakeholders.
- Feedback loops with workers and supervisors.
- Documentation of findings and decisions.

Risk Management

Identifying Risks

Recognize potential risks that could impact the study, such as:

- Resistance from workers or managers.
- Inaccurate or inconsistent data collection.
- Unforeseen operational disruptions.

Mitigation Strategies

Develop strategies to mitigate these risks, including:

- Engaging stakeholders early and often.
- Providing thorough training and support to observers.
- Having contingency plans for data collection and analysis.

"… Are you just going to stand there?"

"And write stuff down. My notes don't reflect on you."

"Uh huh"

"Really, I make it a point in my contract that my observations cannot be used to discipline workers. Anything I observe or you tell me will only be used to change the process."

Chapter 3: Data Collection

Tools and Techniques

Manual Tools

- **Stopwatches**: Traditional tool for time studies. Ensure accurate and consistent timing.
- **Clipboards and Forms**: Standardized forms to record observations systematically.
- **Checklists**: Predefined lists of tasks and motions to streamline data recording. An excellent Standard Operating Procedure for each area is a great starting place.

Digital Tools

- **Video Recordings**: Use cameras to capture tasks for detailed analysis. Ensure high-quality and unobtrusive placement.
- **Software Solutions**: Specialized time and motion study software can automate data collection and analysis.
- **Mobile Apps**: Smartphones and tablets with time study apps can be versatile and convenient for data collection.

Conducting Observations

Establishing a Baseline

- **Initial Observations**: Conduct preliminary observations to understand the process and establish a baseline for comparison.
- **Representative Sampling**: Ensure the data collected is representative of typical operations. This may require observing multiple shifts or different days.

Observation Methods

- **Direct Observation**: Observers watch and record tasks in real-time.
    - Pros: Immediate feedback, context awareness.
    - Cons: Potential observer effect, labor-intensive.
- **Work Sampling**: Randomly sample and record observations at predetermined intervals.
    - Pros: Less intrusive, statistically reliable.
    - Cons: Requires large sample sizes for accuracy.
- **Continuous Observation**: Record all activities over a specific period.

- Pros: Comprehensive data, detailed analysis.
- Cons: Time-consuming, potential for large data volumes.

Recording Data Accurately

- **Standardized Notation**: Use consistent symbols and abbreviations to streamline data entry. A PMTS is a very helpful tool.
- **Detailed Descriptions**: Note the specific actions, tools, and conditions involved in each task. An SOP is a solid place to start, but often does not have quite enough detail for this level of analysis.
- **Time Stamping**: Ensure precise time recording for each observed activity or motion.

Common Pitfalls

Observer Effect

- **Definition**: Changes in worker behavior due to being observed.

- **Mitigation**: Ensure observers are as unobtrusive as possible and conduct longer observation periods to allow workers to acclimate. Allow time for workers to meet and speak with observers and set the stage for process improvement, rather than disciplinary action.

Inconsistent Data Collection Methods

- **Issue**: Variability in data collection can lead to unreliable results.
- **Mitigation**: Train observers thoroughly and use standardized tools and methods. Certification in a PMTS helps with consistency.

Data Entry Errors

- **Issue**: Mistakes in recording or transcribing data can skew results.
- **Mitigation**: Double-check entries, use electronic tools with validation features, and cross-verify with multiple observers.

Data Collection Best Practices

Pilot Testing

- **Purpose**: Conduct a small-scale pilot study to refine your methods and tools before full-scale data collection.
- **Adjustments**: Make necessary adjustments based on pilot results to improve accuracy and efficiency.

Collaboration with Workers

- **Engagement**: Involve workers in the study process to gain their insights and ensure their cooperation.
- **Feedback**: Gather feedback from workers on the feasibility and impact of observations.

Ethical Considerations

- **Transparency**: Inform workers about the purpose and scope of the study. Make it abundantly clear that these observations are about the process, not the worker.
- **Confidentiality**: Ensure the privacy and confidentiality of individual performance data.

Collecting Supplementary Data

Contextual Information

- **Environmental Conditions**: Record factors such as temperature, noise levels, and workspace layout that might affect performance. Understanding which of these environmental conditions affect the process is key; no sense in collecting non-value-added data.
- **Equipment and Tools**: Note the type and condition of equipment and tools used.

Worker Feedback

- **Surveys and Interviews**: Collect qualitative data from workers about challenges, suggestions, and observations. Encourage observers to engage with workers to collect this data, while maintaining productive output.
- **Focus Groups**: Conduct group discussions to gather broader insights and validate findings.

Data Validation and Quality Control

Cross-Verification

- **Multiple Observers**: Use multiple observers to independently collect and compare data for consistency.
- **Inter-Rater Reliability**: Assess the degree of agreement among observers.

Data Cleaning

- **Identifying Outliers**: Detect and investigate outliers that may indicate errors or anomalies.
- **Correcting Errors**: Correct any identified errors and ensure data integrity.

"Another time motion study template?"

"Well, when we made the first one, our scope was completely different."

"Am I going to have to transfer my data?"

"No, I got you. Just use this new one from here on out."

"Until the next one, right?"

Chapter 4: Data Analysis

Organizing Data

Data Entry and Management

- **Data Entry**: Input collected data into a spreadsheet or specialized software. Ensure accuracy by double-checking entries.
- **Data Management**: Organize data systematically using clear labels and categories to facilitate analysis.

Time Sheets, Motion Charts, and Time Motion Worksheets

- **Time Sheets**: Use standardized time sheets to record the duration of each task or motion. Include columns for task description, start and end times, and total duration.
- **Motion Charts**: Visual representations of the sequence and duration of motions within a task. Helpful for identifying inefficiencies and redundancies.

- **Time Motion Worksheet**: A worksheet that allows for the description of each task as well as the PMTS code associated with that task, effectively identifying the motions while assigning a standard time for that motion.

  See Appendix I, II, and III for an example of each.

Identifying Patterns

Bottlenecks

- **Definition**: Points in the process where tasks accumulate, causing delays.
- **Identification**: Look for tasks with significantly longer durations or high variability. Use process mapping to visualize flow and pinpoint bottlenecks.

Redundancies

- **Definition**: Duplicate or unnecessary tasks that do not add value.
- **Identification**: Analyze motion charts and task sequences to identify repetitive actions. Consider worker feedback to validate findings.

Value-Added vs. Non-Value-Added Activities

- **Value-Added**: Activities that directly contribute to the final product or service.
- **Non-Value-Added**: Activities that consume resources without adding value. Examples include waiting, rework, and excessive movement.
- **Analysis**: Categorize tasks and calculate the proportion of time spent on each type. Focus improvement efforts on reducing non-value-added activities.

Data Visualization

Process Mapping

- **Heatmaps or "Spaghetti" Charts**: Visual representations of where the workers are moving in the facility, highlighting high traffic areas or wasteful movement.
- **Swimlane Diagrams**: Flowcharts that separate tasks by roles or departments. Highlight handoffs and potential areas for improvement.

Graphs and Charts

- **Yamazumi Chart**: Also known as a process balance chart. A stacked bar graph that compares task durations and value-added proportions across processes.
- **Pie Charts**: Visualize the proportion of time spent on various activities.

Developing Insights

Identifying Improvement Opportunities

- **Efficiency Gains**: Look for tasks that can be streamlined or automated.
- **Ergonomic Improvements**: Identify motions that can be optimized to reduce worker fatigue and injury.
- **Process Redesign**: Consider reordering or combining tasks to improve flow and reduce delays.

Prioritizing Actions

- **Impact vs. Effort Matrix**: Categorize improvement opportunities based on their potential impact and the effort required to implement them.
- **Quick Wins**: Focus initially on changes that are easy to implement and offer significant benefits.

Reporting Findings

Creating Reports

- **Structure**: Organize reports with clear sections, including an executive summary, methodology, findings, and recommendations.
- **Clarity**: Use straightforward language and avoid jargon. Ensure reports are accessible to all stakeholders.

Presenting Data

- **Visual Aids**: Use graphs, charts, and diagrams to support your findings. Visual aids help convey complex data clearly.
- **Stakeholder Communication**: Tailor presentations to your audience. Focus on key insights and actionable recommendations.

"Wooden platforms? That's the brilliant solution? After all this analysis..."

"I know it sounds simple, but it is the cheapest, fastest, impactful solution. The platforms will reduce walking by 80% and can be built this weekend."

"..."

"An indexing chain conveyor would actually have more of an impact on productivity, but I assumed we wanted improvement THIS year in the realm of the budget I have seen around here."

"I will tell the team they are getting overtime this weekend."

Chapter 5: Implementing Improvements

Developing Solutions

Process Redesign

- **Workflow Optimization:** Analyze the current workflow to identify inefficiencies. Consider rearranging tasks to streamline the process.
- **Elimination of Non-Value-Added Activities:** Identify and eliminate tasks that do not add value to the final product or service.

Ergonomic Improvements

- **Workstation Design:** Optimize workstations to reduce physical strain and improve efficiency.
- **Tools and Equipment:** Ensure that tools and equipment are ergonomically designed and positioned for ease of use.

Automation and Technology

- **Automation**: Implement automation to handle repetitive or time-consuming tasks, freeing up human workers for more complex activities.
- **Technology Integration**: Leverage technology to enhance efficiency and accuracy.

Testing and Validation

Pilot Programs

- **Small-Scale Implementation**: Test proposed improvements on a small scale before a full rollout.
- **Evaluation Criteria**: Define clear criteria for evaluating the success of the pilot program.

Feedback Loops

- **Worker Involvement**: Engage workers in the testing phase to gather their feedback and suggestions. Conduct focus groups or surveys with workers to understand their experiences and insights.

- **Iterative Refinement**: Use feedback to refine and improve the implemented changes. Adjust workstation layout based on worker feedback to further enhance ergonomics and efficiency.

Measuring Success

Post-Implementation Reviews

- **Data Collection**: Continue to collect data on key performance indicators (KPIs) after implementing changes. Track cycle time, throughput, and defect rates before and after implementation.
- **Comparative Analysis**: Compare post-implementation data with baseline data to assess the impact of changes. Use statistical analysis to determine if observed improvements are statistically significant.

Continuous Improvement Cycles

- **PDCA Cycle**: Implement the Plan-Do-Check-Act (PDCA) cycle for ongoing improvement.
    - **Plan**: Identify areas for improvement and develop a plan.

- **Do**: Implement the plan on a small scale.
- **Check**: Evaluate the results of the implementation.
- **Act**: Standardize successful changes and continue to look for new improvement opportunities.

- **Kaizen**: Adopt the continuous improvement philosophy for incremental changes. Encourage workers to suggest small improvements regularly and implement them quickly.

Sustaining Improvements

Standard Operating Procedures (SOPs)

- **Documentation**: Document new processes and procedures to ensure consistency. Create detailed SOPs for new workflows and task sequences.
- **Training**: Provide comprehensive training to workers on new procedures. Conduct training sessions and provide job aids to support the adoption of new practices.

Monitoring and Auditing

- **Regular Audits**: Schedule regular audits to ensure adherence to new processes and identify potential issues. Conduct quarterly audits to review compliance with SOPs and gather worker feedback.
- **Performance Monitoring**: Continuously monitor performance metrics to identify trends and areas for further improvement. Use dashboards and performance scorecards to track KPIs in real-time.

Overcoming Resistance to Change

Change Management Strategies

- **Communication**: Clearly communicate the reasons for change and the benefits to all stakeholders. Hold town hall meetings and send regular updates to keep everyone informed.
- **Involvement**: Involve workers in the change process to gain their buy-in and support. Create cross-functional teams to participate in process redesign and improvement efforts.

Addressing Concerns

- **Listening to Feedback**: Actively listen to concerns and suggestions from workers and address them promptly. Set up a feedback mechanism such as suggestion boxes or an online portal.
- **Support and Resources**: Provide the necessary support and resources to help workers adapt to changes. Offer additional training sessions and provide access to expert assistance.

Celebrating Successes

Recognizing Achievements

- **Public Recognition**: Acknowledge and celebrate the achievements of individuals and teams involved in the improvement efforts.
- **Sharing Success Stories**: Share success stories and case studies to inspire others and build momentum for continuous improvement.

Reinforcing a Culture of Improvement

- **Continuous Learning**: Foster a culture of continuous learning and improvement by encouraging ongoing education and skill development. Offer workshops, seminars, and access to online courses on process improvement and related topics.
- **Empowerment**: Empower workers to identify and implement improvements in their areas of responsibility.

"2020 was wild and challenging and no joke on the 20+ hour days and all-nighters pulled. I was just keeping my nose above the water when [Ted Arehart] showed up. He is like a life raft and our systems and production floor will never be the same. If you are struggling with efficiency and scaling, do yourself a favor and stop the crazy hours + anxiety and get TED Consulting in."

-Bo Nelson, Owner of Thou Mayest Coffee Roasters

"In a matter of weeks, Ted was able to take what applied at Big Business Manufacturing and figure out how to practically apply redesign principles that could be introduced on the micro scale. Ted setup a process for 1 person, 2 person, 3 person and 4 person work teams so the company knew how to scale ahead of time. Companies of all sizes could benefit from TED Consulting."

-Jared Stafos, Head of Business Development at Plexus

Chapter 6: Case Studies

The following pages contain case studies of how time motion studies and the implementation of the discovered improvements can affect various organizations. These studies have unbelievable impact on organizations that already have significant infrastructure. Organizations that already have standard operating procedures, organized workstations, and excellent training programs are perfect for time motion studies. If these are areas that your organization needs improvement, focus on those first before diving into time motion studies.

If you need help, either with time motion studies or getting your organization to the point of needing them, please reach out to me.

Ted Arehart

ted@tedconsulting.net

913-340-2758

Manufacturing

Case Study 1: COVID Coffee Roasting & Packaging (from my own experience)

Background: A small coffee roasting business had to lay off 4 workers due to the facility being unable to support social distancing and new, presumably temporary, safety requirements. Café sales dropped, but online sales skyrocketed, leaving the owner working 20-hour days.

Objective: Reduce overall working time while maintaining productive output.

Methodology:

1. **Initial Analysis**: Observed the 20-hour day, while occasionally asking about which processes were key for quality, effectively knowing nothing about coffee roasting. After 1 day of observation, it became clear that we needed a "spaghetti" diagram.
2. **Data Collection**: Tracked owner's movements throughout next day of production, identifying order and length of time at each station.
3. **Data Analysis**: Drafted factory layout and drew the path which the owner took throughout the day.

Findings:

- **Key Equipment:** Identified crucial machinery: the roasting machine required constant attention, often interrupting other processes.
- **Redundant Motions:** Detected severe redundancy: owner was double handling product and walking between the roaster and packaging areas.
- **Difficult Non-Cyclic Work:** Documented time consuming and physically taxing work: lifting 25-pound buckets overhead to refill machines.

Improvements Implemented:

- **Process Redesign:** Reorganized the packaging process and layout to increase access to roasting machine and reduce handling time. (See drawings below)
- **Ergonomic Adjustments:** Provided ergonomic tools to reduce worker fatigue and improve efficiency.

Results:

- **Cycle Time:** Reduced shift from 20 hours to 8 hours.
- **Throughput:** Remained consistent, even with significant decrease in working time.
- **Worker Satisfaction:** Improved worker satisfaction due to reduced physical strain and clearer processes.

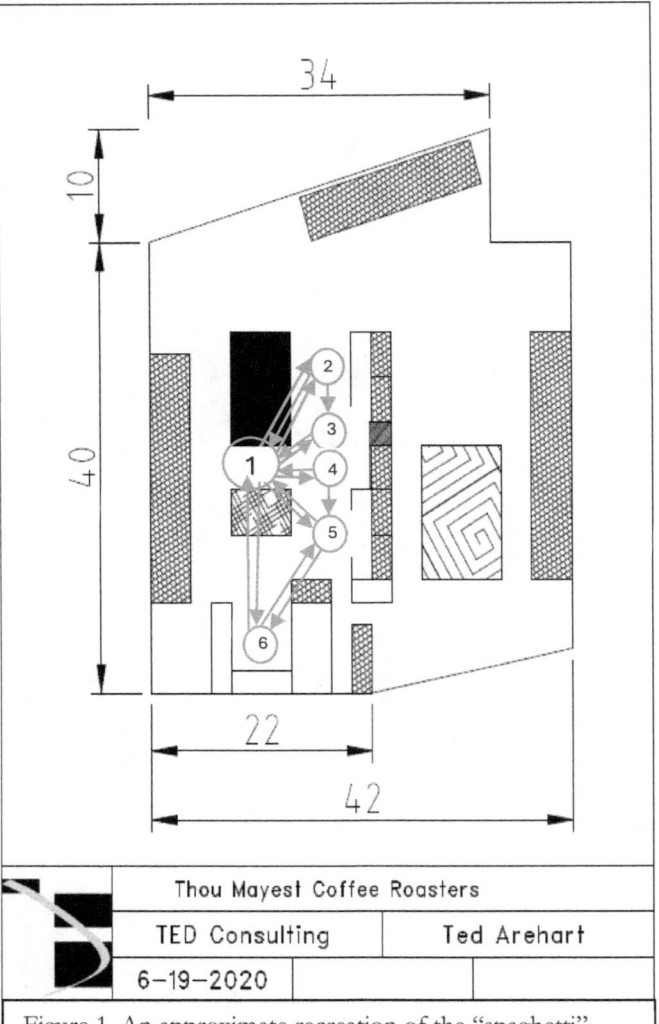

Figure 1. An approximate recreation of the "spaghetti" diagram with original layout. Exterior measurements are in feet. 1 – Roaster, 2 – Bag Labeling, 3 – Weighing, 4 – Sealing, 5 – Bricking, 6 – Boxing.

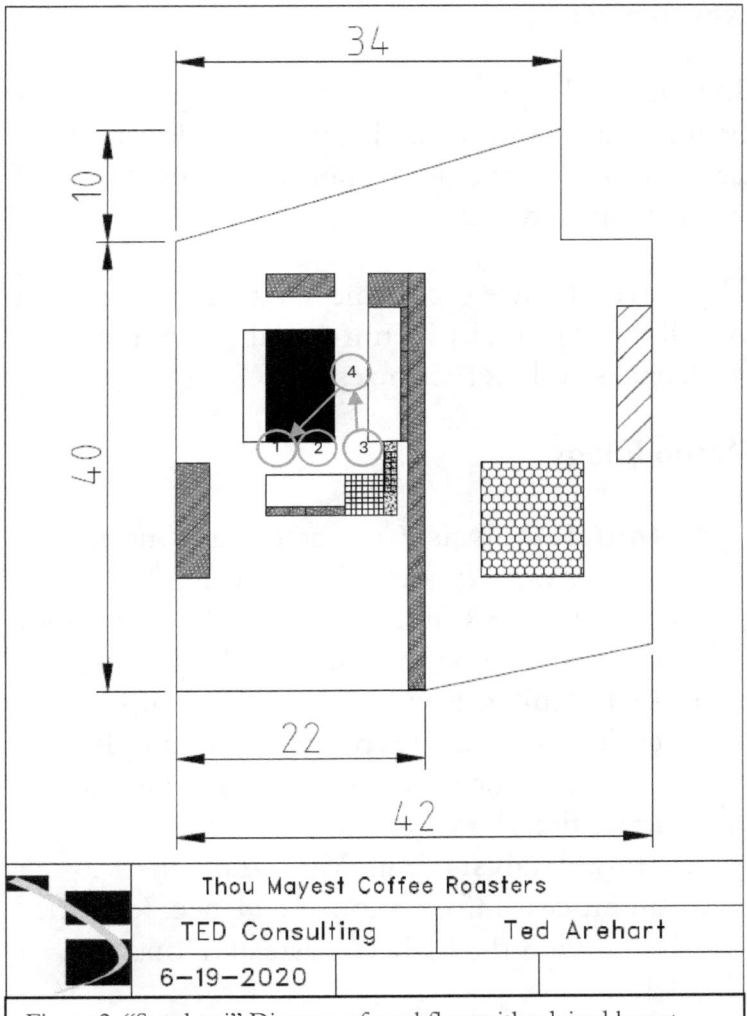

Figure 2. "Spaghetti" Diagram of workflow with advised layout. Exterior dimensions are in feet. 1 – Roaster, 2 – Bag Labeling, 3 – Weighing & Sealing, 4 – Bricking & Boxing. Owner needed to turn or take a single step to get to each station.

Case Study 2: Assembly Line Optimization (AI provided, but shockingly close to one of my experiences)

Background: A mid-sized automotive parts manufacturer noticed significant variability in the assembly line's cycle time, impacting overall productivity and delivery schedules.

Objective: Reduce cycle time variability and increase overall throughput by identifying and eliminating bottlenecks and inefficiencies.

Methodology:

4. **Initial Analysis**: Conducted time motion studies on the assembly line, using MODAPTS® and video recordings to capture detailed data on each step of the process.
5. **Data Collection**: Observed 50 complete cycles to ensure a representative sample. Recorded the time taken for each task and identified the sequence of motions.
6. **Data Analysis**: Built Yamazumi chart to see all process information at a glance. Identified tasks with the highest waste and longest durations.

Findings:

- **Bottlenecks**: Identified two main bottlenecks: a manual inspection step and a sub-assembly process.
- **Redundant Motions**: Detected several redundant motions, particularly in material handling and part retrieval.
- **Worker Feedback**: Gathered insights from workers, highlighting ergonomic issues and suggestions for process improvements.

Improvements Implemented:

- **Automation**: Introduced automated inspection equipment to reduce the manual inspection time and variability.
- **Process Redesign**: Reorganized the sub-assembly process to streamline workflow and reduce handling time.
- **Ergonomic Adjustments**: Improved workstation layout and provided ergonomic tools to reduce worker fatigue and improve efficiency.

Results:

- **Cycle Time**: Reduced average cycle time by 15%.

- **Throughput**: Increased overall throughput by 10%.
- **Worker Satisfaction**: Improved worker satisfaction due to reduced physical strain and clearer processes.

Note: I have done the above project, on multiple occasions in various iterations. An overall throughput of 10% is pretty standard for plants that have had a time motion analysis in the past. I have seen increases of 50% or more for unstudied facilities.

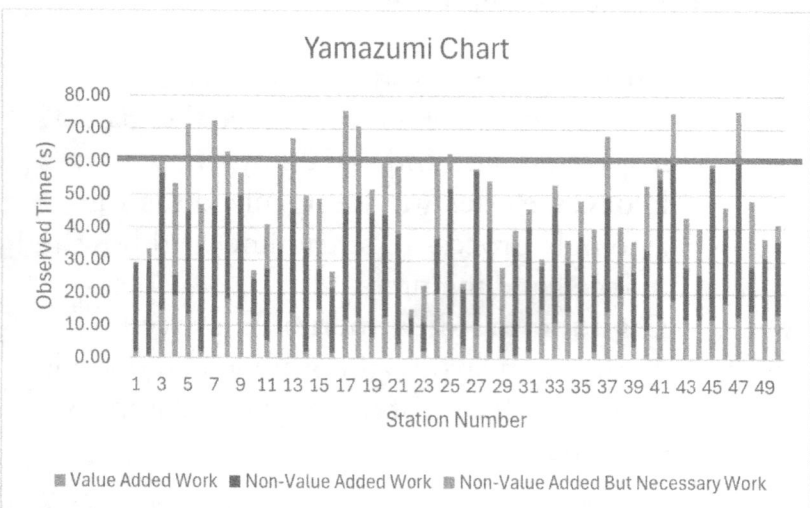

Figure 3. A possible Yamazumi chart of the case study above, quantifying and categorizing the work in each station. The thick line at 60 seconds notes the target cycle time for each station – the goal, if you will. After this is built, the team will determine whether it is best to reduce non-value added work in over utilized station, or move assigned work to a less utilized station.

Healthcare

Case Study 3: Patient Flow Improvement (AI provided)

Background: A large urban hospital experienced delays in patient processing, leading to long wait times in the emergency department (ED) and overall patient dissatisfaction.

Objective: Streamline patient flow from the ED to inpatient wards to reduce wait times and improve patient care quality.

Methodology:

1. **Initial Analysis**: Conducted a time motion study focusing on key stages of patient processing, from triage to admission.
2. **Data Collection**: Used a combination of direct observation and electronic health record (EHR) data to track patient movement and wait times.
3. **Data Analysis**: Created swimlane diagrams to visualize patient flow and identify bottlenecks and delays.

Findings:

- **Bottlenecks**: Identified significant delays during the triage and bed assignment stages.
- **Non-Value-Activities**: Noted considerable time spent on paperwork and repeated information gathering.
- **Resource Allocation**: Discovered imbalances in staff allocation during peak hours, leading to inefficiencies.

Improvements Implemented:

- **Process Standardization**: Standardized triage procedures to ensure consistency and reduce variability.
- **Technology Integration**: Implemented electronic bed management systems to streamline bed assignment and reduce delays.
- **Resource Reallocation**: Adjusted staffing levels to better match peak demand periods.

Results:

- **Wait Times**: Reduced average wait time in the ED by 20%.
- **Patient Satisfaction**: Increased patient satisfaction scores due to reduced wait times and improved care coordination.

- **Operational Efficiency**: Improved overall efficiency, leading to better utilization of resources and reduced overtime costs.

Logistics

Case Study 4: Warehouse Operations (AI provided)

Background: A large e-commerce company faced challenges with order fulfillment accuracy and speed, impacting customer satisfaction and operational costs.

Objective: Improve order fulfillment speed and accuracy by optimizing warehouse layout and processes.

Methodology:

1. **Initial Analysis**: Conducted a time motion study on picking, packing, and shipping processes within the warehouse.
2. **Data Collection**: Used Radio Frequency Identification (RFID) tracking and handheld devices to collect detailed data on movement and task durations.
3. **Data Analysis**: Created heat maps to visualize worker movement and identify areas of congestion and inefficiency.

Findings:

- **Congestion Points:** Identified several congestion points in the picking and packing areas.
- **Inefficient Layout:** Noted that frequently picked items were not optimally located, leading to excessive travel time.
- **Task Redundancies:** Found multiple instances of redundant steps in the packing process.

Improvements Implemented:

- **Warehouse Layout Redesign:** Reorganized the warehouse layout to place high-frequency items closer to the packing stations.
- **Process Automation:** Introduced automated conveyor systems to reduce manual handling and speed up the packing process.
- **Training Programs:** Implemented targeted training programs to improve worker efficiency and reduce errors.

Results:

- **Fulfillment Speed:** Increased order fulfillment speed by 25%.

- **Accuracy**: Improved order accuracy rates by 15%, reducing the number of returns and customer complaints.
- **Cost Savings**: Achieved significant cost savings through reduced labor hours and improved operational efficiency.

Retail

Case Study 5: Checkout Process Enhancement (AI provided)

Background: A national retail chain faced long checkout lines during peak shopping hours, resulting in customer frustration and lost sales.

Objective: Reduce checkout times and enhance customer experience by streamlining the checkout process.

Methodology:

1. **Initial Analysis**: Conducted a time motion study on the checkout process, focusing on cashier and bagging activities.
2. **Data Collection**: Used video recordings and direct observation to capture detailed data on transaction times and customer interactions.

3. **Data Analysis**: Analyzed transaction times, wait times, and customer flow through the checkout lanes.

Findings:

- **Long Transactions**: Identified that complex transactions and payment processing times were major contributors to delays.
- **Queue Management**: Noted inefficiencies in directing customers to open checkout lanes.
- **Employee Training**: Found variations in cashier efficiency due to inconsistent training.

Improvements Implemented:

- **Self-Checkout Stations**: Installed additional self-checkout stations to handle simpler transactions and reduce load on cashiers.
- **Queue Management System**: Implemented an electronic queue management system to direct customers efficiently to open lanes.
- **Standardized Training**: Developed a standardized training program for cashiers to ensure consistency and improve transaction speed.

Results:

- **Checkout Time**: Reduced average checkout time by 30%.
- **Customer Satisfaction**: Increased customer satisfaction scores, with more positive feedback on the shopping experience.
- **Sales**: Saw an increase in sales during peak hours due to reduced wait times and improved customer flow.

Glossary

Automation: The use of technology to perform tasks without human intervention, often implemented to increase efficiency and reduce variability in processes.

Bottleneck: A stage in a process that reduces the overall capacity and flow due to limited throughput.

Cycle Time: The total time required to complete one cycle of a process from start to finish.

Data Collection: The process of gathering information systematically to analyze and use for decision-making.

Kaizen: A Japanese term meaning "continuous improvement," referring to activities that continuously improve all functions and involve all employees.

KPI: Key Process Indicator. A quantifiable metric that indicates success of the organization or process.

Lean Manufacturing: A production methodology that focuses on minimizing waste without sacrificing productivity.

MTM: Methods-Time Measurement. Originally released in 1948 with the goal of eliminating analyst bias, it now exists in a few variations, MTM-1, MTM-2, MTM-SAM, to name a few. Currently maintained and researched by the MTM Association.

MOST: Maynard Operation Sequence Technique. Released in 1972 as an alternative to MTM. Currently maintained and researched by Accenture.

MODAPTS®: Modular Arrangement of Predetermined Time Studies. Developed in 1966 by Chris Heyde to simplify and speed up the process of documenting a time motion study. Currently maintained and researched by the International MODAPTS Association.

Motion Study: The analysis of the movements required to perform a task to identify and eliminate unnecessary motions and improve efficiency.

Pilot Test: A small-scale trial run of a new process or system to evaluate its feasibility and effectiveness before full-scale implementation.

PMTS: Predetermined Time Motion System, a way to quantify motion through set parameters, such as Methods-Time Measurement (MTM), Maynard Operation Sequence Technique (MOST), or Modular Arrangement of Predetermined Time Studies (MODAPTS®)

Queue Management System: A system used to manage customer flow and reduce wait times by directing customers to available service points.

Six Sigma: A data-driven methodology for improving processes by reducing variation and defects.

SOP: Standard Operating Procedure. A document that outlines how to do specific tasks within an organization.

Standard Time: The time established as a benchmark for performing a task under normal working conditions.

Time Study: A technique used to determine the time required to complete a task by measuring and analyzing the time taken for each component of the task.

Utilization: A measure of how effectively resources (such as equipment or labor) are being used.

Work Sampling: A technique for estimating the proportion of time spent on different activities by taking random samples over a period.

Workflow: The sequence of processes through which a piece of work passes from initiation to completion.

Index

Automation, 42, 57, 62, 66
Continuous Improvement, 43
Cost Reduction, 11
Cross-Verification, 32
Data Collection, 26, 29, 30, 43, 50, 56, 59, 61, 63, 66
DMAIC (Define, 16
Frederick Winslow Taylor, 9
Gilbreth, 10
Job Satisfaction, 13
Kaizen, 44, 66
Lean Manufacturing, 10, 15, 67
Motion Study, 9, 14, 15, 67
Pilot Testing, 30
Productivity Improvement, 11
Quality Improvement, 12
Queue Management System, 64, 68
Six Sigma, 16, 68
Standard Time Calculation, 14
Taylor, 9
Time Study, 14, 69
Total Quality Management (TQM), 10
Utilization, 19, 69
Work Sampling, 27, 69
Workflow Optimization, 41

Appendix I: Basic Time Study Sheet

Description of Activity	Time Start	Time End	Total Time for Activity

Appendix II: Basic High Impact Motion Study

Action	Actions/Day	What can we do to reduce this today?	Reduce this week?	Reduce this month?	Reduce this year?
Step					
Move the product without changing the product					
Bend over at the Waist					

Appendix III: Basic Time Motion Study Worksheet (MODAPTS®)

Description	MODAPTS® Code	MODs	Frequency	Time

Notes

Notes

Notes

Notes

Notes

Grid for Workstation Sketch and "Spaghetti" Diagrams

Sketch Page

Grid for Workstation Sketch and "Spaghetti" Diagrams

Sketch Page

Grid for Workstation Sketch and "Spaghetti" Diagrams

Sketch Page

www.ingramcontent.com/pod-product-compliance
Lightning Source LLC
Chambersburg PA
CBHW070118230526
45472CB00004B/1318